Phosphorus NMR of Natural Samples

By Matthew Pasek, Ph.D.

Phosphorus NMR of Natural Samples by Matthew Pasek
Published by Free Radical Consulting
2703 Pemberton Creek Drive, Seffner FL 33584

© 2018 Matthew Pasek

All rights reserved. No portion of this book may be reproduced in any form without permission from the publisher, except as permitted by U.S. copyright law. For permissions contact:

freeradicalconsulting@gmail.com

Contents
1. Why phosphorus 4
2. Fundamentals of NMR 15
3. Solvents 22
4. Concentration and number of scans 24
5. The tools of phosphorus NMR 25
6. The chemical shift 27
7. Phosphorus-phosphorus coupling 33
8. Proton-phosphorus coupling 37
9. Integration and relaxation times 40
10. Solid state phosphorus 44
11. Example spectra 45

1. Why phosphorus

The element phosphorus is a key nutrient in many ecosystems. This is due in large part to its prominent role in biochemistry. Phosphorus—along with sugars such as ribose and deoxyribose—make up the backbone of the nucleic acids RNA and DNA, where genetic information is stored. In addition, the metabolic molecule adenosine triphosphate (ATP) stores the chemical energy of life, releasing it when an organism needs to do work. It is estimated that the average organism makes and breaks its body mass in ATP each day. Phospholipids also make up cell membranes, separating organisms from their outside environment.

In most of these molecules (but not all) phosphorus is surrounded by four oxygen atoms. This arrangement is termed phosphate. Phosphates can also be organic esters (organophosphates) with one or two organic functional groups (respectively termed organophosphate monoester or diester), or they can be dimers/trimers/polymers of phosphate (pyrophosphate, triphosphate, polyphosphate). Phosphate on its own is sometimes termed orthophosphate. Organophosphate triesters are not found in nature to the best of my knowledge.

In addition to phosphates there are a number of less common phosphorus ions that occasionally occur in nature. In all of these molecules, one or more oxygen atom is replaced by a heteroatom, such as hydrogen, carbon, phosphorus, nitrogen, sulfur, or fluorine. Each of these has been shown, or proposed, to have been important on the surface of the earth at some point in the earth's history.

The replacement of one oxygen atom with hydrogen yields the ion phosphite (HPO_3^{2-}). The hydrogen atom is directly bonded to the phosphorus atom in this molecule, and is not acidic under typical conditions. This molecule has been shown to have been present on the early earth [1] and may have played a much larger role in prebiotic chemistry than has been predicted in the past. Like phosphate, bonding an organic compound through one of the oxygen atoms gives a phosphite ester. No triesters are possible under typical conditions, and diesters are as uncommon as organophosphate triesters. Phosphite may form a dimer called diphosphate or pyrophosphite, and this molecule has been proposed as a primary prebiotic phosphorus source[2]. No trimers or polymers are known to occur naturally, as these

Changing the phosphorus atom for a carbon atom or organic radical gives the phosphonate compounds. These compounds have turned out to be ubiquitous in biology, and many have antibiotic

---

[1] https://www.nature.com/articles/s41467-018-03835-3
[2] http://www.mdpi.com/2075-1729/3/3/386

properties. They also are environmentally important, controlling methane production in the ocean[3]. Some of the most common organophosphonates are 2-aminoethylphosphonate, and methylphosphonate.

Replacing an oxygen atom with a sulfur atom gives the thiophosphates. Thiophosphates are the only known substitute for phosphate in DNA, as some organisms make thiophosphorylated [4]DNA in response to oxidants as a defense. Like many of these compounds, thiophosphates form esters, including monoesters and diesters (the latter of which includes thiophosphate DNA).

Replacing one oxygen with a phosphorus atom gives the unusual molecule hypophosphate. Hypophosphate has yet to be discovered in natural samples, though it is highly likely to be present based on its production by the corrosion [5]of certain meteorite minerals in the laboratory. No organic esters of hypophosphate are known. However, phosphite and phosphate can form a dimer called isohypophosphate when dried down that has a formula and average oxidation state similar to hypophosphate. Hypophosphate and isohypophosphate can be easily distinguished by NMR.

---

[3] https://www.nature.com/articles/ngeo234
[4] https://www.nature.com/articles/nchembio.2007.39
[5] http://www.pnas.org/content/105/3/853

If an oxygen is replaced with a nitrogen or amine group, then you have a class of compounds called the amidophosphates. Monoamidophosphate and diamidophosphate have one and two oxygen atoms replaced with -NH$_2$ groups, respectively. The diamidophosphate ion was recently identified[6] as a plausible molecule that could have caused the formation of organophosphates on the early earth. In addition, it is possible to replace an oxygen atom in the phosphite ion with an -NH$_2$ group, giving monoamidophosphite.

Another reduced phosphorus species is the hypophosphite ion (H$_2$PO$_2^-$) where phosphorus has the nominal oxidation state of +1. This compound is relatively unstable (compared to phosphite and many of the organophosphorus compounds) and oxidizes to phosphate over weeks to months. It has been found in the hindguts of termites (where the wood is broken down to food)[7]. Associated with hypophosphite are the organophosphinates, where two P-O bonds are replaced with P-C bonds, and these compounds include naturally occurring antibiotics.

The final group of alternatives to phosphate are the fluorophosphates, where one P-O bond is replaced with a P-F bond. One mineral, termed

---
[6] https://www.nature.com/articles/nchem.2878
[7] https://academic.oup.com/chromsci/article/49/8/573/379097

bobdownsite, appears to bear this moiety[8], though NMR proof was not given (and NMR would be conclusive for this ion).

Note that all the above are the names of the ions as they are commonly known by practicing phosphorus chemists. However, the official names of each of these compounds is different as specified by the International Union of Pure and Applied Chemists (IUPAC).

---

[8] http://www.canmin.org/content/49/4/1065.abstract

Phosphate

$$\begin{array}{c} O \\ \| \\ {}^-O-P-O^- \\ | \\ O^- \end{array}$$

Orthophosphate diester

$$\xi-CH_2-O-\overset{\overset{O}{\|}}{\underset{\underset{O^-}{|}}{P}}-O-CH_2-\xi$$

Polyphosphate (diphosphate or pyrophosphate)

$$-O-\underset{\underset{O^-}{|}}{\overset{\overset{O}{\|}}{P}}-O-\underset{\underset{O^-}{|}}{\overset{\overset{O}{\|}}{P}}-O^-$$

Phosphite or H-phosphonate

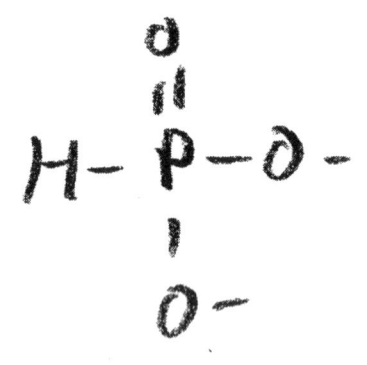

$$H-\underset{\underset{O^-}{|}}{\overset{\overset{O}{\|}}{P}}-O^-$$

Diphosphate or pyrophosphite

$$\text{H}-\underset{\underset{\text{O}^-}{|}}{\overset{\overset{\text{O}}{\|}}{\text{P}}}-\text{O}-\underset{\underset{\text{O}^-}{|}}{\overset{\overset{\text{O}}{\|}}{\text{P}}}-\text{H}$$

Phosphonate (2-aminoethylphosphonate)

$$\text{NH}_2-\text{CH}_2-\text{CH}_2-\underset{\underset{\text{O}^-}{|}}{\overset{\overset{\text{O}}{\|}}{\text{P}}}-\text{O}^-$$

Thiophosphate

$$\begin{array}{c} S \\ \| \\ {}^-O - P - O^- \\ | \\ O^- \end{array}$$

Hypophosphate

$$\begin{array}{c} O \quad\quad O \\ \| \quad\quad \| \\ HO - P - P - O^- \\ | \quad\quad | \\ O^- \quad O^- \end{array}$$

Isohypophosphate

$$H-\underset{\underset{O^-}{|}}{\overset{\overset{O}{\|}}{P}}-O-\underset{\underset{O^-}{|}}{\overset{\overset{O}{\|}}{P}}-O^-$$

Diamidophosphate

$$H_2N-\underset{\underset{O^-}{|}}{\overset{\overset{O}{\|}}{P}}-NH_2$$

Hypophosphite (H-phosphinate)

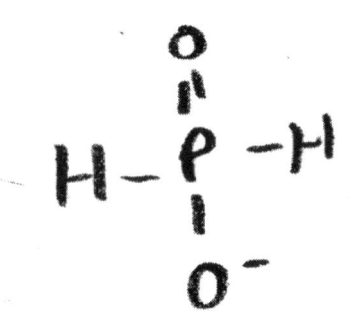

$$\text{H}-\overset{\overset{\text{O}}{\|}}{\underset{\underset{\text{O}^-}{|}}{\text{P}}}-\text{H}$$

Fluorophosphate

$$^-\text{O}-\overset{\overset{\text{O}}{\|}}{\underset{\underset{\text{O}^-}{|}}{\text{P}}}-\text{F}$$

2. Fundamentals of NMR

Nuclear Magnetic Resonance (NMR) Spectroscopy is a tool that uses the natural nuclear spin of many atoms to identify molecular structures. The spin of an atom's nucleus is one of its fundamental quantum properties, and hence NMR has nothing to do with nuclear fission or fusion, as your organic chemistry professor will certainly make clear. Additionally, there are entire courses and careers built on NMR. A more thorough overview of NMR, including the theory and quantum mechanics behind it, can be found in other texts. What follows is a basic (even cartoony) discussion of NMR.

Not all atomic nuclei are NMR sensitive. In order to have a <u>nuclear spin</u>, an atom's nucleus must have an odd number of protons, neutrons, or both. The most commonly used elements in NMR all have spins of ½, which results from an even/odd combination of the protons/neutrons or vice versa. Hydrogen, carbon-13, fluorine and phosphorus all have spins of ½. Atomic nuclei that have a non-zero nuclear spins can be visualized as tiny magnets. As the nucleus spins, the axis of its spin also <u>precesses</u>, much as a spinning top (or perhaps more relevantly, a fidget spinner) will wobble regularly. Nuclides that have spins greater than ½ are termed quadrupolar and often have very broad peaks, and are often too broad to be analyzed using a typical NMR.

When a nucleus with a spin is placed in a strong magnet, there then exists an energy difference between aligning the nucleus's spin with the magnetic field, and aligning opposite of the magnetic field. The latter is higher energy. The energy <u>difference</u> is very slight, but can be strengthened by increasing the magnetic field strength. To this end, the main chamber of an NMR spectrometer is a giant magnet, typically composed of a superconductor in an ultra-cold fluid (usually liquid helium) that provides an extremely high magnetic field. NMR magnets have field strengths on the order of several tesla (T). In contrast, the earth's magnetic field is about 30 µT, and a magnet for a fridge is a few mT. The strongest magnets generally available—known as neodymium magnets—have field strengths of about 1 T on their surface.

When placed in such a strong <u>magnet</u>, isotopes of a given element will resonate at a frequency proportionate to the magnetic field strength. To this end, most NMRs in chemistry labs are given as "200 MHz" or "400 MHz". These numbers refer to the resonant frequency of hydrogen (^{1}H) in the NMR. In other words, in a 400 MHz NMR, a typical hydrogen atom will resonate when hit with a radio wave with a frequency of 400 MHz (FM radio is about 100 MHz, AM radio is about 1 MHz).

NMR irradiates nuclei placed within a strong magnetic field with a specific radio frequency, usually a <u>pulse</u>. This adds enough energy to make the number of nuclei whose spins are oppositely aligned with the magnetic field equivalent to the number that are aligned. When the radio pulse ceases, these nuclei then <u>release a radio frequency</u> back as a few of the oppositely aligned nuclei release energy and return to the aligned state. This frequency is very close to the frequency that the nucleus was irradiated with, but usually differs in frequency by a few parts per million and up to 100s of parts per million in some cases. The deviation of this frequency from the incident frequency is termed a molecules' chemical shift.

The signal strength of the frequency given back (termed free induction decay or FID) as the nuclei relax back to the aligned state is dependent on a few factors: temperature, the strength of the magnet, the abundance of the element in the sample, the abundance of the NMR-sensitive isotope in question, time between pulses, and the susceptibility of the isotope to resonance.

The signal received back (the FID) may be a mix of different signals, and so it must be processed to separate different peaks. This is done through Fourier transform, which calculates the specific frequencies within the FID and produces a spectrum of these peaks.

This book covers the specifics of phosphorus NMR. Phosphorus is one of the easiest elements to analyze by NMR (only H and F are easier, in general), because it has a spin of +½ and because the element is monoisotopic: effectively all phosphorus that occurs in nature is ^{31}P, the NMR sensitive nuclide. In comparison, ^{13}C is slightly more sensitive, but is much less common in the environment, so its analysis by NMR is slow and less easy.

The frequency an NMR operates when analyzing 3^{1P} is equal to 0.404807 × the proton frequency. So a 300 MHz NMR analyzes ^{31}P at 121.43-121.44 MHz, and a 500 MHz NMR analyzes at about 202.4 MHz.

Spin

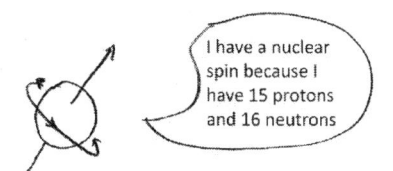

I have a nuclear spin because I have 15 protons and 16 neutrons

Precession

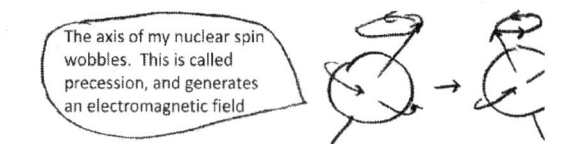

The axis of my nuclear spin wobbles. This is called precession, and generates an electromagnetic field

Field

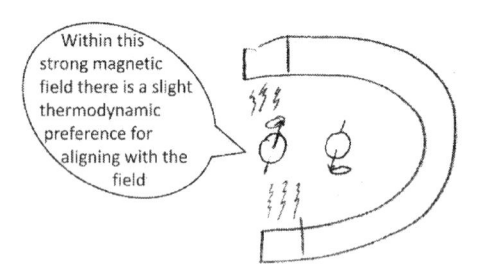

Within this strong magnetic field there is a slight thermodynamic preference for aligning with the field

Equilibrium

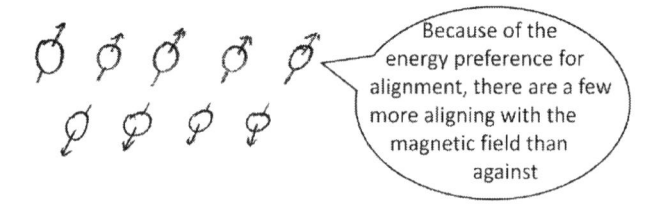

Because of the energy preference for alignment, there are a few more aligning with the magnetic field than against

Pulse

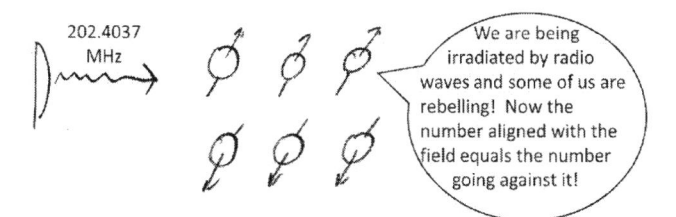

202.4037 MHz

"We are being irradiated by radio waves and some of us are rebelling! Now the number aligned with the field equals the number going against it!"

Relaxation

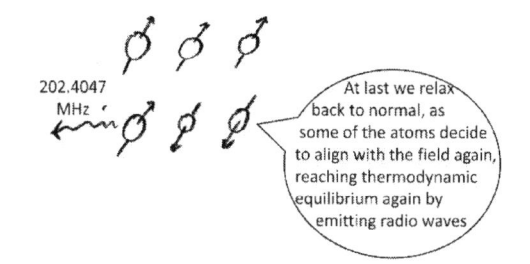

202.4047 MHz

"At last we relax back to normal, as some of the atoms decide to align with the field again, reaching thermodynamic equilibrium again by emitting radio waves"

3. Solvents

The solvent in which you dissolve your solvent is a key consideration when preforming NMR experiments. However, fortunately for naturally-occurring phosphorus compounds the options are limited. The cheapest and generally most effective is D_2O or deuterated water. The compounds that you will be looking at should either be water-soluble, or should be made soluble by adjusting the solution chemistry to dissolve the phases within. Dimethyl sulfoxide or DMSO may also work, if water causes problems in your experiments, but it is much more expensive and somewhat less versatile.

As an example, the dissolution of phosphate minerals such as the apatite group—$Ca_5(PO_4)_3(OH,F,Cl)$—occurs best at low pH or with using chelating agents such as acetic acid or EDTA. I have found that EDTA at ~0.05 M works well for many phosphate minerals and other phases, though minerals usually do not completely dissolve. This may not be an issue for most experiments, but in case it is, you may find that by decreasing the amount of mineral you attempt to dissolve you may be able to promote its complete dissolution.

When analyzing natural phosphorus samples by NMR, there are a few caveats and concerns of which you should take note. The first is the role ionic strength in sample analysis. If you have a solution with a high ionic strength (> 100 mM), then tuning the NMR to detect a given element may be problematic, especially on more sensitive NMR machines.

If you don't remember the formula for ionic strength, it is:

$$\frac{1}{2} \text{SUM}(c_i z_i^2) = \text{I.S.}$$

Ionic strength is half of the sum of the concentration of each ion in your solution multiplied by its charge squared. Remember that charge is pH-dependent (the phosphate ion is -3 only above a pH of 12). Too high an ionic strength will produce a noisy or otherwise bad spectrum, even if you have a concentrated solution of the ion of interest. Ways around this include using less salt (obviously), or by using an ion exchange column. An ionic exchange column works because for every Mg^{2+} you exchange for two Na^+, you halve the ionic strength (because of z^2).

The other common issue with solutions in NMR is the precipitation, adsorption, or complexation of phosphate salts after dissolution. As an example, in many of my experiments, the slow oxidation of Fe^{2+} to Fe^{3+} by air results in the removal of phosphate as the phosphate binds more strongly to Fe^{3+} than Fe^{2+}. This can be dealt with by using chelating agents (for Mg^{2+} and Ca^{2+}), or by precipitating problematic ions by adjusting pH (Fe^{3+} reacts under basic solutions to form $Fe(III)(OH)_x(O)_y$ compounds).

4. Concentration and number of scans

NMR is not a sensitive technique. Unlike various forms of MS and HPLC, most of the time the sampling limits are on the order of millimolar. Phosphorus NMR tends to be a long-term NMR experiment. In my experience, most natural samples require at least 256 scans to get a good signal (standards are fine to run at 16-64 scans), and that's under the best conditions. A handy chart is shown below, for concentrations of compounds of interest in molarity (moles/L).

expected concentration	typical numbers of scans
0.1	64
0.01	256
0.001	1000
0.0001	6000
0.00001	32000

Due to limits of sensitivity, attempting to detect P compounds with concentrations of 10 µM and less is typically a futile endeavor. Note also that this is for a single experiment, so doing a set of proton-decoupled and proton-coupled scans will take twice as long.

5. The tools of phosphorus NMR

Phosphorus NMR centers around two main tools: chemical shift and coupling constants. Peak integration is also a highly valuable tool under certain circumstances, but in general integration is less useful for P than for organic chemistry proton NMR (where 3 Hs tends to imply a methyl group). There are occasions where integrations can be useful for species identification, but generally separate peaks in ^{31}P NMR are separate molecular species.

The chemical shift is calculated from the wavelength emitted by phosphorus nuclei as they relax back into the equilibrium state. This wavelength typically differs from the pulsed wavelength by a few kHz. Relative to the pulsed wavelength this small variation is on the order of a few parts per million relative to a standard. For phosphorus, the standard is 85% H_3PO_4, though in practice it is uncommon to add this standard to your sample, as this standard is quite acidic. An enclosed capillary tube may be placed in your NMR tube as a reference, but others add compounds such as trimethylphosphate to a solution as a standard.

The second major tool of phosphorus NMR is J-coupling constants. Phosphorus is an NMR-sensitive nuclide, and if it happens to occur in close proximity to other NMR-sensitive nuclides (most commonly H or a different P atom), then the nuclides will cause emitted frequencies to split.

Tools summary

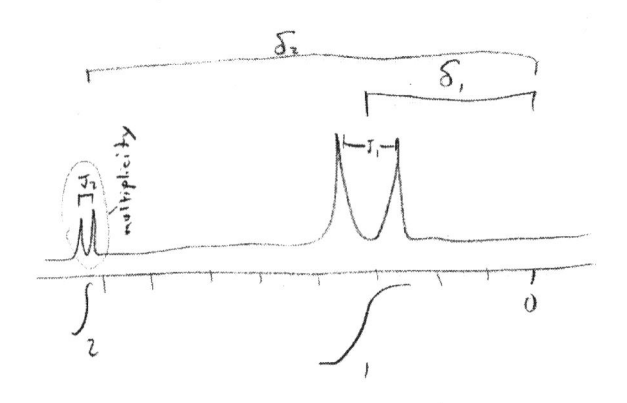

6. The chemical shift

The chemical environment of an atom's nucleus causes a slight variation in the frequency a nuclide emits when it relaxes to the "ground" state. These variations are usually not large, and are generally measured in ppm, that is, parts per million variations from a standard frequency. If a sample with phosphorus is placed in a 500 MHz NMR (irradiated at ~202.4037 MHz), it might emit a radio wave with a frequency of 202.4047 MHz. This is a difference of $0.001/202.4037 \times 10^6 = \sim+5$ ppm. Thus, if a sample has P atoms with this chemistry, there will be a peak at 5 ppm on the ^{31}P NMR spectrum.

For phosphorus, chemical shift can be a powerful tool for the identification of speciation. Since phosphorus is such a good element to analyze by NMR, its peaks tend to be narrow when optimized and separate quite well. Peak overlap is usually not a problem, except under certain conditions.

The location of a given peak is indicative of its chemical identity. Nuclides are said to be "deshielded" or "downfield" when they have a spin with a higher absolute number than some reference (so hypophosphate at +14 ppm is downfield of phosphate at +4 ppm). The opposite is "shielded" or "upfield", and the terms derive from how well the NMR-sensitive nuclides feel the radio pulse in the NMR. Those that feel it more are said to be deshielded, and hence have a higher energy and a higher ppm when they relax back to equilibrium.

The chemical shift of phosphorus molecules is dependent on a number of things: pH, polymerization, solvent, organic constituents and the regiochemistry of the organic radical, in addition to the specific tweaking and setup of the analyzing NMR.

You have control over the solvent you use, and if you're analyzing natural P compounds, you're likely using D_2O as your locking solvent. The specifics of the NMR (its tunable frequency, magnetic field strength, and probe type) you likely do not have as much control over, unless you use NMR extensively as a tool. To this end you will likely need a reference compound with which to compare your NMR spectra. As stated in the prior chapter, the reference compound is by default phosphoric acid, H_3PO_4 (85% in water). This is set to 0 ppm, and all P compounds are referenced to it. It is not practical to add this compound to your experiments as the compound may be highly reactive, or may swamp out the small peaks you're looking for, hence you will probably need to run this standard occasionally while using the NMR to correct for drifts in the chemical shift.

Hypothetically you can use any compound as a standard as long as it's been referenced to H_3PO_4. My personal favorite is Na_2HPO_3 (~0.01 M in D_2O) as it is a compound I used frequently in my research, and it showed up consistently at +4.15 ppm. If you see a deviation in the expected shift on your reference compound, you will need to correct for it. In general, this correction can be applied arithmetically. Subtract the deviation of your reference compound from the peak locations of all of the peaks in your spectrum to have these match your prior spectra (so, if your H_3PO_4 reference comes in at 0.50 ppm, and you see a peak at 8.34 ppm, adjust it upfield to 7.84 ppm).

  An often less controllable cause of shift in ppm is from changes in pH. At specific pHs, compounds will move either upfield or downfield as acids lose protons. As a general rule, most compounds move downfield as the solvent becomes more basic (pH increases). A study provided by Yoza and others[9] demonstrated that some compounds shifted by as much as +5 ppm as pH increased from 3 to ~11.

  This shift is due to the loss of the shielding effect of protons. With increasing pH, acidic protons are lost and the P nuclides are deshielded. In fact, if you were to perform a P NMR scan at a variety of pHs you could estimate the pKas of the P molecule. Indeed, if the P molecule has a pKa greater than 11, then if you raise the pH of your solution above 11 it will continue to deshield. For

---

[9] https://link.springer.com/article/10.1007/BF00325563

instance, phosphate begins at about 0 ppm and moves up to ~6.75 ppm at a pH of 13.

For this reason, you should measure the pH of your sample prior to analysis by NMR. This can be done using simple pH paper, or by using more complex tools. Since the pH may shift by as much as ~7 ppm from 0-14 pH, using peak position to identify a molecule must take this effect into account. As a quick approximation, you may assume a peak will change by 2 ppm for each proton lost (+2) or gained (-2).

As an example, consider the case of phosphate. H_3PO_4 is by definition at 0 ppm, $H_2PO_4^-$ occurs at 2 ppm, HPO_4^{2-} occurs at 4 ppm, and PO_4^{3-} is about 6.75 ppm. This is consistent with deshielding with loss of protons. However, if the phosphate is esterified, then diesters (R-O-P(O)$_2$-O-R') occur at about 0 to 2 ppm (DNA is close to 0 ppm), as the phosphate is shielded by the addition of organics, and hence does not suffer from deshielding with increased pH. A monoester ($ROPO_3^{2-}$) is similarly located at around 4 ppm at high pH, again because it is not as deshielded as the associated phosphate. So as a second rule, the addition of an organic as an ester tends to increase shielding of the ^{31}P nucleus. As a second quick approximation, each peak will change its position by -1 ppm for each organic ester, relative to the unesterified P molecule.

The following table should help with identification of a P molecule based on its chemical shift. As stated at the beginning of this book, we are looking primarily at 4 coordinate, naturally occurring phosphate compounds. Synthetic compounds such as phosphines, phosphonium salts, and phosphoranes remain well outside of my familiarity and have hence been excluded. The peaks I have observed have all lied between -30 and +50 ppm, with the majority between -25 and +36 ppm. Note also that these peak positions should be independent of NMR magnet strength. Phosphite in basic conditions should lie a 4 ppm, in both a 300 and 600 MHz NMR.

Compound	Shift at pH 7	Name
HPO_4^{2-}	4	phosphate
HPO_3^{2-}	4	phosphite
$ROPO_3^{2-}$	4	orthophosphate monoester
$HP_2O_7^{3-}$	-7	pyrophosphate
$HP_3O_{10}^{4-}$	-8	triphosphate
$HP_3O_{10}^{4-}$	-18	triphosphate
$HP_3O_9^{2-}$	-21	trimetaphosphate
$H_2PO_2^-$	8	hypophosphite
$HP_2O_6^{3-}$	12	hypophosphate
$ROP(O)_2H^-$	3	phosphite monoester
$ROP(O)_2OR'$	0	orthophosphate diester
$RO-PO_2-OR$	17	cyclic phosphate diesters
SPO_3H^{2-}	35	thiophosphate
RPO_3^{2-}	10-35	phosphonate
$H_2NPO_3^{2-}$	5	monoamidophosphate
FPO_3^{2-}	1	fluorophosphate
$(H_2N)_2PO_2^-$	8	diamidophosphate
$H_2P_2O_6^{2-}$	-4	isohypophosphite
$H_2P_2O_5^{2-}$	-4	pyrophosphite
$RPHO_2^-$	20-30	organophosphinates

7. Phosphorus-phosphorus coupling

When a nuclide is in close proximity to another nuclide that is also resonant when placed in the magnet, then the nuclides will interact and split each other with respect to their magnetic resonance frequency. Another way of saying this is that if magnet A is close to magnet B, then magnet A can either have its magnetic field strength added to or subtracted from its proximity to magnet B. This effect is called coupling.

There are two common cases where this occurs with phosphorus. We examine first the coupling when phosphorus interacts with another phosphorus nucleus.

Coupling of two sensitive nuclei occurs when the nuclei are three bonds or less apart. For phosphorus, this means P-P, P-X-P, and P-X-Y-P bonds are the ones that will exhibit coupling. Coupling only happens if the chemical environment around the two nuclei is different, though. So if the molecule exhibits symmetry, then you would not expect to see any coupling. In the case of pyrophosphate: $O_3P-O-PO_3$, the two phosphorus nuclei do not split each other. However, in the case of adenosine diphosphate, $O_3P-O-PO_3$-adenosine, the two phosphorus atoms would split each other, since the two phosphorus atoms no longer are in identical chemical environments.

If two phosphorus atoms are more than three bonds apart, you should not expect to see any coupling of the two atoms.

The coupling of two nuclei results in "splitting" of an NMR peak, and the terms are often used interchangeably, like I've done above. There are two important pieces of information to extract from this. The first is the splitting pattern. If a P nucleus is close to <u>one</u> other P nucleus that is chemically different, the its peak is split into a doublet. In contrast, if the P (let's call it P#1) nucleus is close to <u>two</u> others that are chemically different from P#1 (but not chemically different from each other), then its peak is a triplet. The splitting pattern is called the peak's multiplicity. With phosphorus, you should expect to see doublets, triplets, and occasionally a doublet of doublets meaning a P atom next to two different P atoms. This latter case occurs when P#1 is near to P#2 and P#3, and P#2 and P#3 are not chemically identical.

The second piece of information extracted from peak splitting is something called the J-coupling constant. The J-coupling constant is calculated by measuring the peak position in ppm of the <u>separate peaks of a multiplet</u>. This is then multiplied by the frequency, in MHz, of the NMR. So for a 400 MHz NMR operating for phosphorus at 161.9 MHz, a 0.01 ppm difference in peak location of a doublet corresponds to a J coupling constant of 16.2 Hz (the M in MHz cancels out the ppm). This can be stated as a formula:

NMR MHz for H × 0.405 × peak difference in ppm = J-coupling constant

The J-coupling constant for a 2-bond P-P interaction is typically between 15-20 Hz. When doing ^{31}P NMR, you will see this coupling even in proton-decoupled spectra. This splitting is indicative of triphosphate, tetraphosphate, polyphosphate, and the organic diphosphates and triphosphates (e.g., ATP and ADP), and the somewhat more obscure molecule isohypophosphate, which occurs when phosphate and phosphite link through an oxygen atom. There are other occurrences, too, but in natural samples these are the ones I would expect the most.

There do exist other varieties of phosphorus-X coupling. The next chapter focuses on P-H coupling, but P-F, P-^{13}C, and P-^{15}N coupling are all feasible, though much less common in nature. These types of couplings can be seen in well-designed experiments, often done as part of syntheses. Of these, a P-F bond has a J-coupling constant of about 870 Hz. I am unfamiliar with the other two coupling constants (though P-^{15}N appears to be about 10-50 Hz)[10], and neither should be expected as common in nature.

It is worth noting that J-coupling constants are constant across NMR magnet strengths (so they're the same in a 300 MHz and a 600 MHz), but the splitting distances of the peaks change, as a consequence of the formula given above.

J Coupling Constant

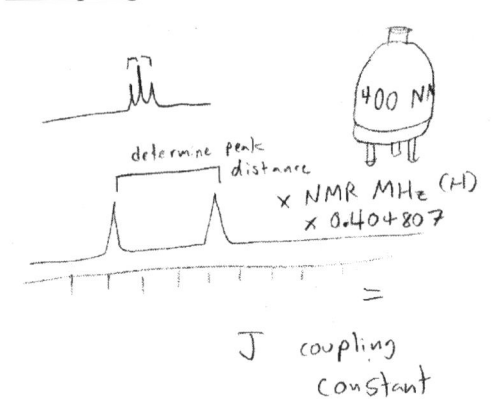

$$\frac{\text{determine peak distance} \times \text{NMR MHz (H)} \times 0.404807}{} = J \text{ coupling constant}$$

(400 NMR)

---

[10] Gombler, W., Kinas, R. W., & Stec, W. J. (1983). 31P–15N Coupling Constants and 15N/14N Isotope Effects on 31P NMR Chemical Shifts of 2-Phenylamino-2-oxo (-thioxo,-selenoxo)-4-methyl-1, 3, 2-dioxaphosphorinanes and Related Compounds. *Zeitschrift für Naturforschung B*, *38*(7), 815-818.

8. Proton-phosphorus coupling

More generally useful than phosphorus-phosphorus coupling is proton-phosphorus coupling. This is because there are more H-P intramolecular interactions than P-P intramolecular interactions. In H-P coupling, the bond distance of a phosphorus atom from a hydrogen atom can be determined by the size of the J-coupling constant. Additionally, the coupling should be observable in H-NMR allowing a double check of this chemistry, assuming the H NMR spectrum is not swamped out by other species (for instance H_2O in D_2O).

Most ^{31}P NMR scans have options of being done in either "coupled" or "decoupled" mode. I recommend if at all possible doing both. In decoupled mode, the hydrogen atoms are constantly irradiated and hence do not split the frequency of any neighboring phosphorus atoms. As a result, ^{31}P NMR peaks are strong singlets (aside from any P-P or P-X couplings beyond P-H). In coupled mode this does not occur, and hence any hydrogen atoms within three bonds of a phosphorus atom in a molecule will couple with each other. A useful experiment is to run ^{31}P NMR in decoupled mode for 100-10000 scans, then repeat it in coupled mode. Then you can overlay the two spectra, and can observe any P-H couplings clearly. All split peaks should be symmetrical about the decoupled peak.

For H-P interactions, the J-coupling constants can be diagnostic of the ions of interest. For instance, the H-P J coupling constant of phosphite is 565 Hz, where the H and P are bonded to each other. If you observe a peak in decoupled mode around 4 ppm, then in coupled mode observe this turn into a doublet with a coupling constant of about 560-570 Hz, you clearly have phosphite (HPO_3^{2-}) in the sample. If in contrast you have a peak at about 8 ppm that forms a triplet with a J constant of about 520 Hz when proton coupling is turned on, then you clearly have hypophosphite ($H_2PO_2^-$). Alternatively, if you have a peak at 4 ppm that splits into a narrow triplet with a J-coupling constant of 7 Hz, then you have an orthophosphate monoester (H_2C-O-P). The H-P J-coupling constant varies slightly with pH (a few Hz at most), but, coupled with multiplicity and peak position (in ppm) you usually have enough data to make a quick identification of the species, at least with respect to phosphorus moiety. Phosphorus NMR is not specific enough to tell you the difference between ethylphosphate and 1-propylphosphate, but provides a good starting point.

Note that it is not terribly common to see a 2-bond H-P coupling. The phosphonates may show a 2-bond coupling (methylphosphonate, for instance, has a clear 2-bond H-P coupling), but sometimes this coupling is absent, especially when dealing with organophosphinates (e.g., $H-P(O)_2-$

C(H)(OH)CH₃). In these compounds the 2-bond H-P coupling is zero.

A handy chart of H-P J coupling constants vs. bond distance is provided below.

Bond distance	Coupling Constant (Hz)
H-P (1)	250-900
H-X-P (2)	15-25
H-X-Y-P (3)	5-10

9. Integration and relaxation times

The final piece of information extractable from NMR is the relative abundance of specific species. This is found through integration under the NMR peaks. Such a process is usually software-driven, and need not be performed by hand. Since most ^{31}P NMR peaks have similar shapes, peak height can also be used to estimate relative abundances, though peak area is certainly preferable.

Effectively, the premise is that the amount of nuclides with a given magnetization release the same number of photons as they relax. The result should be that the peak area is directly proportional to the molarity of each constituent.

If done properly and carefully (with standards) you can use NMR to get a rough estimate of the concentration of a given species. In some of my prior work[11] I used the signal to noise ratio vs. the square root of number of scans to arrive at a relationship between concentration and NMR peak height. Your mileage may vary, and it is not guaranteed that this will work across all varieties of NMR.

11

https://www.sciencedirect.com/science/article/pii/S001670370700021X

There is a major caveat with assuming peak area is proportional to abundance in ^{31}P NMR, however. This is that the phosphorus atoms may relax on slower timescales than the NMR analysis will permit. The time it takes from the radio pulse to reestablish equilibrium in the NMR chamber is called the relaxation time. For phosphorus this will range from less than a second up to ~20 seconds. If you pulse again prior to this 20 seconds passing, then it is possible that some molecules have not fully relaxed and thus would be undercounted by NMR. Most ^{31}P NMR experiments are set up with shorter than 20 seconds of relaxation time (e.g., 1 to 3 seconds) and thus they do not produce accurate peaks-to-concentration relations. You can improve the ^{31}P NMR quantitation if the relaxation time in the NMR program is set to 30 seconds, but that will greatly lengthen your NMR analysis time. If you are not time-limited that is not a problem, but if you can only reserve 30 minute blocks, then that is unlikely to be feasible. Usually you should perform one NMR experiment for identification, and then if quantification is important, perform a quantitative NMR subsequently.

That said, there are two caveats to this major caveat. First, the relative peak area ratios usually are quantitative within a single molecule. The best example of this is triphosphate, which has a doublet for the terminal phosphates and a triplet for the interior phosphate. The peak area of the doublet is almost always twice that of the triplet. This is because the relaxation time of both types of phosphorus within the molecule is close enough that they will give the expected 2:1 ratio independent of relaxation time.

The other major caveat is that the presence of certain ions can greatly decrease the relaxation time, especially iron. Iron in solution decreases ^{31}P NMR rates to one tenth or even more compared to the iron-free solution. This is because iron can pull off extra magnetism from phosphorus quickly. As a result, the quantitation of ^{31}P in solutions with iron is usually better than those without.

As an example, the below are the results of one of my experiments with various P compounds, all with 1 mM initial concentration (+/- 1%). The relaxation effect (as well as something called the nuclear Overhauser effect or NOE) alter the determined areas under each curve:

Compound	Peak area
Isopropylphosphonic acid	1.08
Ethylphosphonic acid	0.95
Methylphosphonic acid	0.98
Hypophosphate	0.84
Phosphite	1.03
Orthophosphate	1.16
Pyrophosphate	1.14
Acetylphosphonic acid	0.9
Hypophosphite	0.88

10. Solid state phosphorus

Solid state NMR is a tool that identifies the speciation of NMR-sensitive elements in solids. Many of the tools described above (integration, peak location, J-coupling) are applicable to solid state NMR (especially peak location for speciation identification), but there are many more issues. In most cases the peaks within solid state NMR are extremely broad, making identification more challenging. I recommend that, if possible, you attempt to dissolve a portion of a sample in water/D_2O for liquid state NMR as liquid NMR is much easier to interpret than solid state NMR, because the peaks are sharper, and the coupling is much more obvious.

11. Example spectra

Several example NMR spectra are given below. These are some of what you might expect to see in NMR, for selected compounds. These are on a 500 MHz NMR and are "synthetic", so you may see something that varies from this. These are shown proton coupled, though phosphite and diphosphite also show proton decoupled.

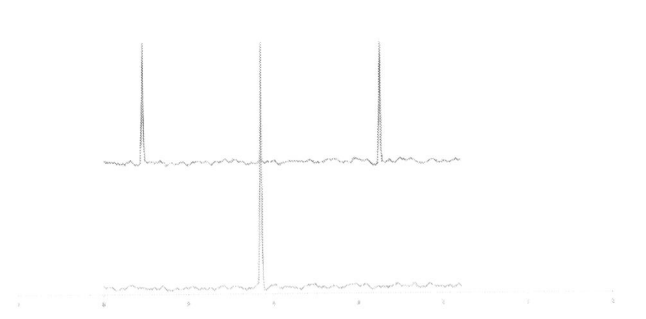

Phosphate spectrum

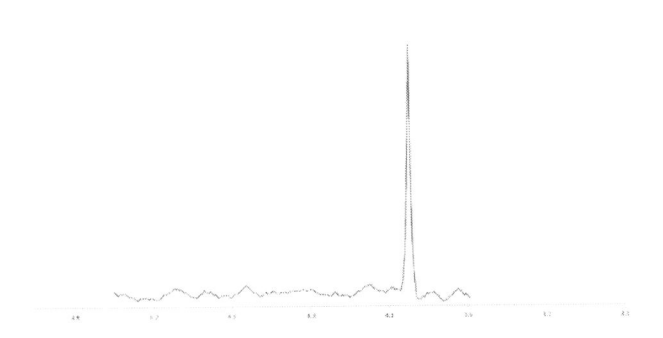

Phosphite spectrum

Thiophosphate spectrum

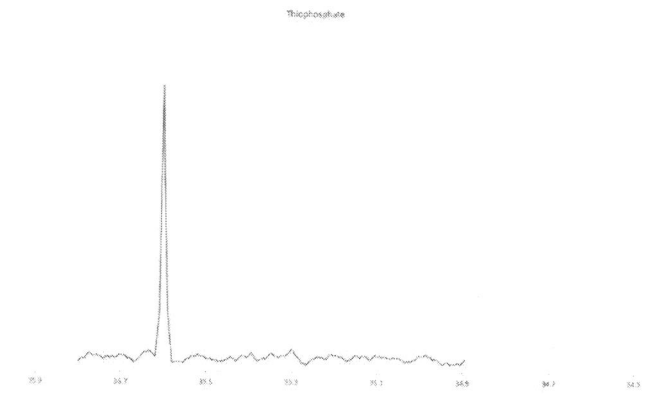

5'-adenosine monophosphate spectrum

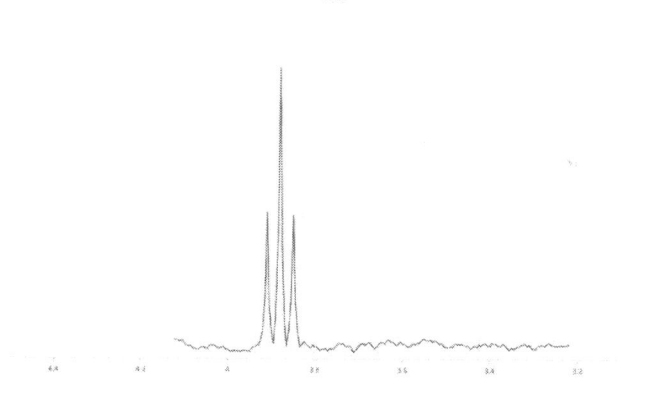

3'-adenosine monophosphate spectrum

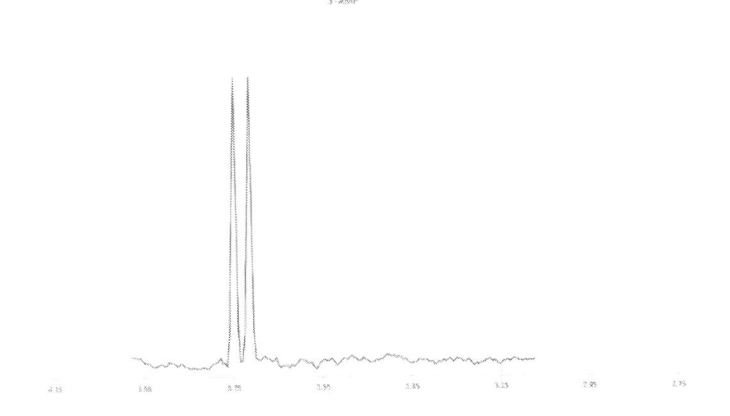

1,2-Cyclic glycerol phosphate spectrum

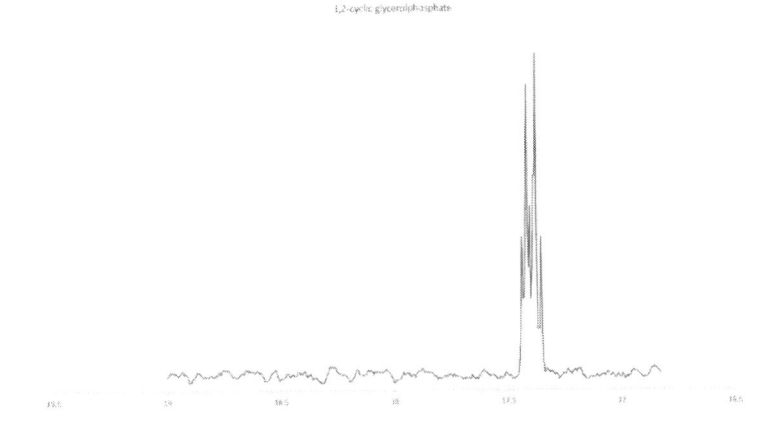

Methylphosphonate spectrum

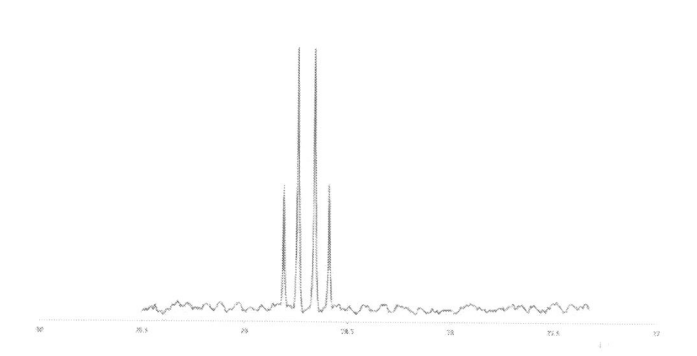

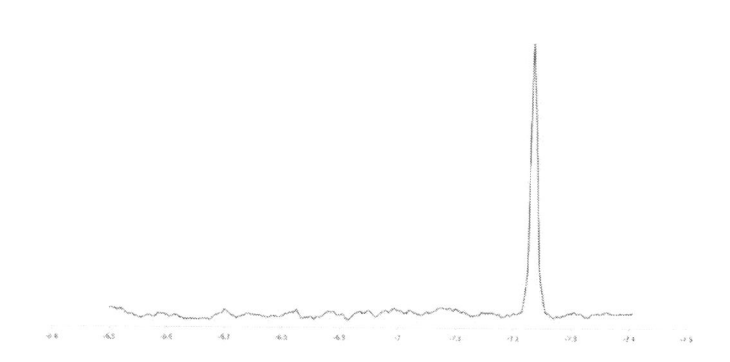

Pyrophosphate spectrum

Cyclic trimetaphosphate spectrum

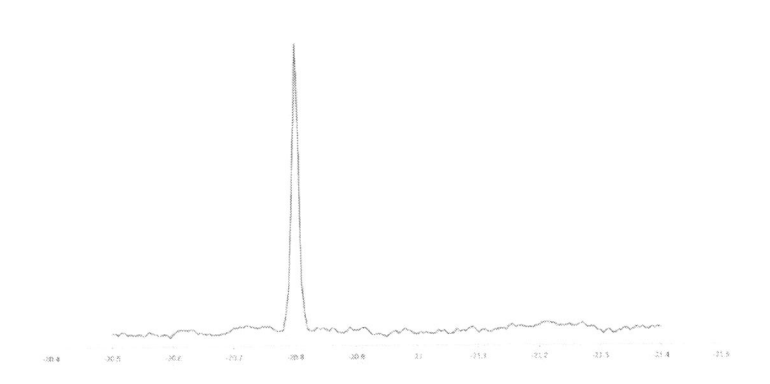

Triphosphate spectrum

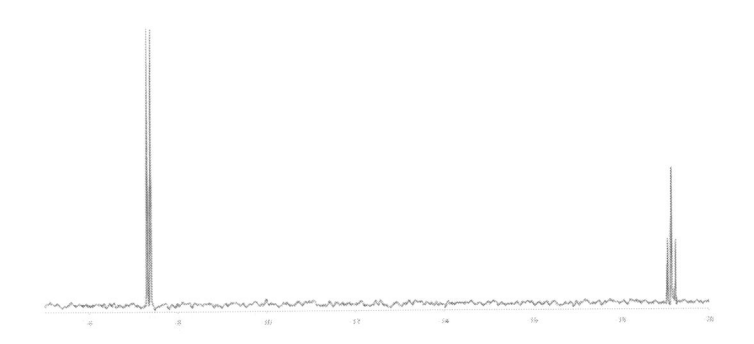

Diphosphite spectrum

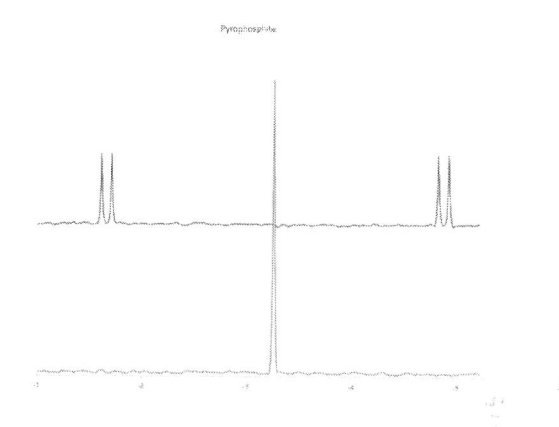

Made in United States
Orlando, FL
31 January 2022